SOFT CINEMA

navigating the database

Lev Manovich | Andreas Kratky

Introductions

Cinema and Software
Lev Manovich

The Future Was Then
Sheldon Brown

The Maturity of New Media
Jeffrey Shaw

Films

TEXAS

MISSION TO EARTH

ABSENCES

Lev Manovich | Andreas Kratky
Soft Cinema: Navigating the Database

The MIT Press, 2005
Massachusetts Institute of Technology
Cambridge, Massachusetts 02142
http://mitpress.mit.edu

© The MIT Press, 2005

ISBN 0-262-13456-X

Produced with the assistance of:

BALTIC	The Centre for Contemporary Art, Gateshead, UK
CAL-IT (2)	(California Institute for Telecommunications and Information Technology), San Diego and Irvine, USA
CRCA	(Center for Research in Computing and the Arts), University of California - San Diego, USA
RIXC	(The Centre for New Media Culture), Riga, Latvia
ZKM	(Center for Art and Media), Karlsruhe, Germany

Cinema and Software

The twentieth century cinema 'machine' was born at the intersection of the two key technologies of the industrial era: the engine that drives movement and the electricity that powers it. While an engine moves film inside the projector at uniform speed, the electric bulb makes possible the projection of the film image on to the screen.

The use of an engine makes the cinema machine similar to an industrial factory organized around an assembly line. A factory produces identical objects that are coming from the assembly line at regular intervals. Similarly, a film projector spits out images, all the same size, all moving at the same speed. As a result, the flickering irregularity typical of the moving image toys of the nineteenth century is replaced by the standardization and uniformity typical of all industrial products.

Cinema also reflects the logic of the industrial era in another way. Ford's assembly line, introduced in 1913, relied on the separation of the production process into a set of repetitive, sequential, simple activities. Similarly, cinema replaced previous modes of visual narration with a sequential narrative and an assembly line of shots that appear on the screen one at a time.

Given that the logic of the cinema machine was closely linked to the logic of the industrial age, what kind of cinema can we expect in the information age? Rather than waiting for this new cinema to appear, the Soft Cinema project generates new cinema forms using the key technology of the information society - a digital computer.

As I have already explained, the logic of twentieth century cinema was not directly connected to the operation of an engine but instead reflected the industrial logic of mass production, which the engine made possible. Similarly, the Soft Cinema project is interested not in the digital computer per se, but rather in the new structures of production and consumption enabled by computing.

drawings by Lev Manovich, 1981–1991

Lev Manovich

Manovich was born in Moscow where he studied painting, architecture, computer programming and semiotics. After having practiced fine arts for a number of years, he immigrated to New York in 1981. This geographical move catalyzed a logical shift in his interests from the still image and physical 3D space to the moving image, virtual space and the use of digital computers. He worked professionally in the field of 3D computer animation from 1984 to 1992 while completing an M.A. in Experimental Psychology and a Ph.D. in Visual and Cultural Studies.

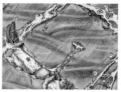

Since the early 1990s, his work has combined art practice, theory, lecturing and teaching. As a visual artist, his projects that investigate the possibilities of post-computer cinema have been presented by, among others, ZKM, the Walker Art Center, KIASMA, Centre Pompidou, and the ICA, London. His publications include *The Language of New Media* and *Tekstura: Russian Essays on Visual Culture*, as well as many articles that have been published in over 30 countries. Manovich is a Professor in the Visual Arts Department at the University of California, San Diego, where he teaches courses in new media art and theory.

The Future Was Then

In the future, cinema will be: `void dead ( );` `int eractive ( );` `char wet ( );` `struct complex {double immersive; double ubiquitous;};` `typdef struct complex subversive implanted organic continuous.`

There are many other functions, classes and variable types by which to declare what it is that will create the foundations of the future of cinema. Soft Cinema provokes speculation on this, but it does so not by positing a future cinema but by enacting a present cinema. After I watch the works that constitute Soft Cinema, the normative cinema of my time feels nostalgic.

New media art is science fiction. It operates by extrapolating cultural vectors that are technologically inflected. There is good sci-fi and bad sci-fi, and bad sci-fi that can be seen as good with the right attitude. The making of good sci-fi is grounded in a clarity about the direction of cultural vectors. It is grounded in possibilities that extend out from the actualities of transformation, not from pure fantasy. These actualities catalyze the work with the vitality of consequence – thus the sci-fi of new media art becomes the expression of the particular moment of a culture and not a speculated future.

I have always thought that Lev only does the simplest things in his work. What he does is state the obvious. Soft Cinema is obviously the cinema of our moment. It's just that no one has done it until now.

SHELDON BROWN
Director of Center for Research in Computing and the Arts (CRCA)
University of California, San Diego

The Maturity of New Media

One of the benefits of making art in the early days of new media was that new media operated outside of the cultural mainstream. As a result, exterior interests and pressures were few and the exigencies of the work itself were free to drive the creative process. But this fecund seclusion also had its drawbacks, for there were few opportunities to exhibit works produced and even fewer occasions on which anything intelligible was written about them.

For some time new media art practice suffered from this lack of an adequate critical commentary, while the commentary that did exist typically ranged from techno-rapture to an even more livid techno-mysticism. Most problematic of all was an emerging movement of cultural theorists who did not have a language to express the actual processes of new media art creation. Notwithstanding the socio-political value of their work, this circumstance allowed these theorists to superimpose theoretical constructs that transformed and deformed the identity of the works way beyond their makers' recognition and intentions.

Lev Manovich's The MIT Press publication *Language of New Media* was a turning point in regard to articulating the actual processes of digital creation. With his book a coherent and revelatory interpretation of new media appeared and it was written by a practicing artist in the field. In other words, it was written by an analyst whose theoretical position was founded on, and could be verified by, the nature of the practice itself.

Lev is cognizant of the technological underpinnings of the new media environment – the properties that inspire, facilitate, constrain and frustrate the artist in equal measure. In the same way that a good painting demonstrates how a specific handling of brush strokes can constitute a pictorial achievement, so the successful media artwork demonstrates a precise physical and conceptual transformation of its materials, as opposed to a lesser work that is typically subsumed by the materials.

The comprehensive understanding that is manifested in Lev's theoretical texts has now come to inform his art practice as well. Soft Cinema is the return of theory out of practice, to the further formation of practice informed by theory. It is a higher level of practice that is born from a personal process of meditation on the 'language of new media'.

I was happy to have had the opportunity to invite Lev, as artist in residence at the ZKM Institute for Visual Media, to work on the Soft Cinema project together with Andreas Kratky, and then in 2002 to be able to present it as one of the benchmark highlights of the Future Cinema exhibition that I curated together with Peter Weibel. And I am delighted that Soft Cinema has now developed into this excellent DVD publication, for it will now have the opportunity to edify and entertain an even larger public and take a prominent place in the history of new media culture alongside Lev's inimitable writings.

JEFFREY SHAW
Director, iCinema (Centre for Interactive Cinema Research)
University of New South Wales, Sydney, Australia

 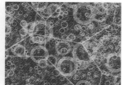

details from MISSION TO EARTH

Database / Sampling

The first database comprises 425 short video clips selected from footage that I have shot in various locations over a number of years. Extending the genre of a 'city film' from the 1920s, the database is constructed to capture the iconography of a 'global city'.

The second is a music database created by the composer George Lewis as a parallel to the video database. It consists of samples taken from his own archive of sounds – his own version of a 'global layer' – as well as from his earlier compositions. The two databases are correlated because they use the same parameter – 'type of space' – to arrange their samples. (In other words, both video clips and sound files are described using the same spatial categories: 'city view', 'space with screen', 'private interior', 'public interior', 'object', 'working with screen'.)

Along with Lewis's database, the film soundtrack uses tracks from the CD *The Quick and The Dead* by DJ Spooky and Scanner. The CD represents the meeting between different 'database imaginaries' of these two outstanding artists. DJ Spooky brings numerous music traditions, genres, and sound cultures into a single vast sound space through sampling. His music can be thought of as a systematic traversal of a multi-dimensional sound database in every possible direction. Equally versatile and prolific in his output, Scanner often generates his sound databases using a variety of procedures and logics for recording sound in all kinds of environments. In the words of the artist, "in some ways my work is concerned with capturing, hunting sound from many inaccessible spaces and bringing it out, whether it's the private phone conversations I find in an airspace that proved more public than anyone thought, or location recordings from the restricted access sites which my art projects take me to" (from February 2003 interview, online at www.scannerdot.com). Therefore, if the Texas video database reflects visible and spatial characteristics of the 'global city,' *The Quick and The Dead* captures both its public sound and its less visible communication dimensions: "floating above the city: waves, frequency bursts, packets of distilled information distributed throughout the spectrum of all communications devices" (DJ Spooky, from "Web Notes for *The Quick and the Dead*" at www.djspooky.com).

Andreas Kratky

Born in Berlin, Kratky studied visual communication, fine arts and philosophy in Berlin and Paris. His art projects include *Postkarten für die Hauptstadt*, Berlin; *Berliner - Tonale Portraits*, Berlin; and *mondophrenetic*, Brussels (collaboration with INCIDENT VZW). Kratky is responsible for media design and co-direction on the award winning DVDs *That's Kyogen* and *Bleeding Through - Layers of Los Angeles 1920-1986* (both published by ZKM), as well as a number of other multimedia publications. He has also collaborated on research projects dealing with information visualization and interface design at Karlsruhe and Manchester Universities. Since 1998 Kratky has worked at ZKM | Center for Art and Media, and in 2002 he was appointed head of ZKM's Multimedia Studio. Since mid 2004 Kratky has been working as an independent media artist. He is currently designing and co-directing several DVD projects with the University of Southern California, Los Angeles; Humboldt Universität, Berlin; and Université de Paris 1 Panthéon-Sorbonne.

details from ABSENCES

How can we represent the subjective experience of a person living in a global information society? If daily interaction with volumes of data and numerous messages is part of our new 'data-subjectivity', how can we visualize this subjectivity in new ways using new media – without resorting to already normalized modernist techniques of montage, surrealism and the absurd?

Today many places look and feel like composites made up from different layers: 'traditional', 'global', 'capitalist', 'post-communist', etc. How to represent the typical modern experience of living 'between layers' – between the past and the present, between East and West, between there and here?

Texas aims to address these questions by using a number of specific techniques. The film exists at the intersection of a number of databases, each of which is structurally organized in the same way and each of which can be thought of as a portrait of a contemporary 'global layer.' (In other words, each database is a different set of samples from the same territory.) When the film is playing, the Soft Cinema software selects samples from these sets and mixes them in real time.

TEXAS

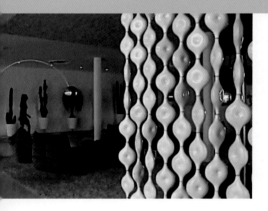

CREDITS

The original version of *Texas* was created for the 2002 Soft Cinema installation
that was commissioned by ZKM Center for Art and Media for the exhibition
Future Cinema: Cinematic Imaginary after Film. This DVD presents the 2004 version of the film, which has new narration, music, sound design, and additional
graphics.

[Lev Manovich | narrative, videography, animations, editing rules] [Schoenerwis-
sen/OfCD | Berlin|visualization] [DJ Spooky | Scanner | New York, London | music
from CD *The Quick and The Dead*] [George Lewis | New York | music]
[Kelly Richardson|Newcastle| media management] [David Ung | San Diego | nar-
rative graphics] [Iryna Zinchenko | San Diego | sound editing] [Lee Anne Schmitt |
Los Angeles | voice over]

Visualization

While the Soft Cinema Project uses a database as the 'engine' that generates the movies, we should also think of the database as a new representational form in its own right. Accordingly, we asked Schoenerwissen / Office for Computational Design to translate our video database into a new visual representation. The resulting dynamic visualization of 425 video clips represents each clip as a small square, while the human-ascribed subjective descriptions of the clips appear to float on the screen. Additionally - since it is the key parameter in *Texas* - the visualization appropriately foregrounds 'geo location' by having each of the squares orbit around a point that represents the city or country in which the original video clip was shot.

TEXAS video database visualization
by Schoenerwissen

TEXAS video database (a partial view)

Proust / Google

If for Proust and Freud modern subjectivity was organized around a narrative - the search to identify that singular, private and unique childhood experience which had defined the identity of the adult - subjectivity in the information society may function more like a search engine. In *Texas* this search engine endlessly mines through a database that contains personal experiences, brand images, and fragments of publicly shared knowledge. The operation is revealed when the characters in the story communicate: they semi-randomly jump from one retrieved 'record' to another - similar to the way in which the Soft Cinema software retrieves and plays the clips from the video database. While the jumps are always triggered by something - a question in the conversation, the taste of a drink or meal - the retrieved records are only loosely connected to the outside world and to each other.

Database Aesthetics

The editing of the video database in *Texas* follows the same poetics of record retrieval, i.e. weak connections between the displayed records and abrupt shifts from one record to the next. The clips that the software selects to play one after another are always connected on some dimension - geographical location, type of movement in the shot, type of location, and so on - but the dimension can change randomly from sequence to sequence. In addition, in contrast to a traditional film, there are no dissolves or cross-fades. Instead one screen layout is instantly replaced by another. In a nutshell, the 'hard' aesthetic of a traditional narrative is replaced by the 'soft' aesthetic of a database narrative.

Finally, the content of *Texas* addresses the contemporary subjective experience of living 'between the layers' in yet another way. The film belongs to the series of Soft Cinema editions that I have called *GUI* (Global User Interface). Each story in the *GUI* series occurs in a different location: Texas, Hamburg, Kiev, Mongolia. The narratives take place in the present, which has been put through a light science fiction filter. (However, since in writing them I followed the principle that they can only take place in locations that I have never visited as an adult, perhaps they are more accurate than I can imagine.)

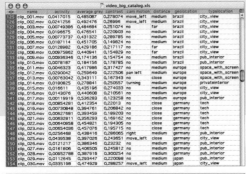

Logging clips into the database

Editing with Soft Cinema software

Between Narrative and a Search Engine

Each video clip in the *Texas* database is described by 10 parameters that specify where the video was shot, the nature of its subject matter, its average brightness and contrast, the type of space, the degree and type of camera motion, and so on. These parameters are used by the software in assembling the movies. Starting with a particular clip, the software finds other clips that are similar to it on some dimensions. This is similar to the way in which we use web search engines such as Google. When Google returns a number of results for a particular search term we can say that all these results are connected on a few dimensions: the search term, language, domains, etc.

In the case of *Texas* what you see on screen while the movie is playing are multiple sequences generated in a similar manner. Each sequence is the result of a particular search through the Soft Cinema database. Each is perhaps equivalent to a 'scene' in a normal film, while a series of such searches ('scenes') becomes equivalent to a traditional film. Film editing is thereby reinterpreted as the search through the database. Consequently it is possible to describe *Texas* as a media object that exists 'between narrative and a search engine'.

Scanner /music

British artist Robin Rimbaud (aka Scanner) creates absorbing, multi-layered soundscapes that twist technology in unconventional ways. From his early controversial work using found mobile phone conversations through to his current focus on trawling the hidden noise of the modern metropolis, his restless explorations of the experimental terrain have won him international admiration from, among others, Bjork and Stockhausen. Scanner is committed to working with cutting edge practitioners and has collaborated with artists from every imaginable genre: musicians Bryan Ferry and Laurie Anderson, The Royal Ballet and Random Dance companies, composers Michael Nyman and Luc Ferrari, and artists Mike Kelley and Derek Jarman. In addition to producing compositions and audio CDs, his diverse body of work includes soundtracks for films, performances, radio, and site-specific intermedia installations. He has performed and created works in many of the world's most prestigious spaces including SFMOMA, the Hayward Gallery, Centre Pompidou, the Corcoran Gallery of Art, Tate Modern, and the Royal Opera House, London.

▶ www.scannerdot.com

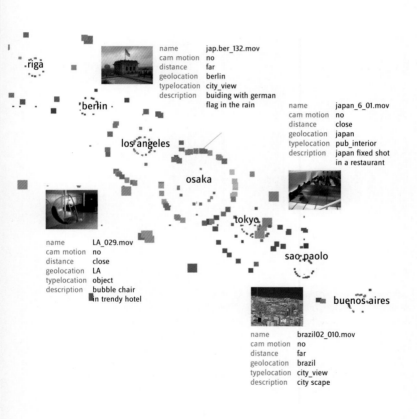

riga

name   jap.ber_132.mov
cam motion no
distance   far
geolocation berlin
typelocation city_view
description buiding with german
             flag in the rain

berlin

name   japan_6_01.mov
cam motion no
distance   close
geolocation japan
typelocation pub_interior
description japan fixed shot
             in a restaurant

los angeles

osaka

tokyo

name   LA_029.mov
cam motion no
distance   close
geolocation LA
typelocation object
description bubble chair
             in trendy hotel

sao paolo

buenos aires

name   brazil02_010.mov
cam motion no
distance   far
geolocation brazil
typelocation city_view
description city scape

selected clips from TEXAS database
with their keywords (superimposed
over database visualization)

details from TEXAS

Schoenerwissen / OfCD / video database visualization

Schoenerwissen / Office for Computational Design was founded in 1998 by Marcus Hauer and Anne Pascual. SW/OfCD develops software and carries out research in a broad range of areas, including visual network applications, data mapping systems, and information visualization. In designing dynamic and open processes that implement temporal and spatial parameters SW/OfCD looks for new models of representation and aims to make the non-perceived elements of data processing visible to a general user. Hauer and Pascual studied at the Academy of Media Arts, Cologne. Their project *Minitasking*, a visualization of the *Gnutella* peer-to-peer network, won both an Award of Distinction in the net excellence category of the Prix Ars Electronica 2002 and the Transmediale Software Award in 2003.

George Lewis / music

George Lewis is an improviser-trombonist, composer, and computer/installation artist. He studied composition with Muhal Richard Abrams at the AACM School of Music and trombone with Dean Hey. The recipient of a MacArthur Fellowship in 2002, a Cal Arts/Alpert Award in the Arts in 1999, and numerous fellowships from the National Endowment for the Arts, Lewis has explored electronic and computer music, computer-based multimedia installations, text-sound works, and notated forms. Lewis' work as composer, improviser, performer and interpreter is documented on more than 120 recordings. His oral history is archived in Yale University's collection *Major Figures in American Music*, while his articles on music, experimental video, visual art, and cultural studies have appeared in many scholarly journals and edited volumes. The University of Chicago Press will publish Lewis' forthcoming book titled *Power Stronger Than Itself: The Association for the Advancement of Creative Musicians.*

details from TEXAS

Earth

main sources: video shot in London, Berlin,
Rio de Janeiro, Buenos Aires, Sweden (1999–2003)

1 Tokyo
2|3 Los Angeles
4 Newcastle (UK)

DJ Spooky /music

Paul D. Miller is a conceptual artist, writer and musician working in NYC. His writings have appeared in, among others, *The Village Voice*, *The Source*, *Artforum*, *Raygun*, *Rap Pages*, and *Paper Magazine*. He was the first Editor-At-Large for *Artbyte: the Magazine of Digital Culture* and he is a co-publisher, along with the legendary African American downtown poet Steve Canon, of *A Gathering of the Tribes* – a periodical dedicated to new works by writers from a multi-cultural context. Miller's book *Rhythm Science* (The MIT Press, 2004) was named one of the best books of the year by The Guardian and Publishers Weekly.

Miller's art projects have appeared in a wide variety of contexts, including the Whitney Biennial; The Venice Architecture Biennale; The Ludwig Museum, Cologne; Kunsthalle, Vienna; and The Andy Warhol Museum, Pittsburgh. Miller is most well known under the moniker of his 'constructed persona' as 'DJ Spooky that Subliminal Kid'. As DJ Spooky he uses digitally created music as a form of post-modern sculpture and he has recorded a huge volume of music, including the score for the Cannes and Sundance award winning film *Slam*. He has also collaborated with a wide variety of renowned musicians and composers, including, among others, Iannis Xenakis, Ryuichi Sakamoto, Butch Morris, Kool Keith (aka Doctor Octagon), Killa Priest from Wu-Tang Clan, Yoko Ono, and Thurston Moore from Sonic Youth.

▶ www.djspooky.com

Frame from a clip
used in TEXAS (shot in Riga)

Inga is an alien who comes to Earth from Alpha-1, a planet that is about twenty years behind Earth culturally and technologically. *Mission to Earth* is an allegory of both the Cold War era and of the contemporary immigrant experience that is so frequently the norm for inhabitants of 'global cities'. The film reminds us that, while hybrid identity is often celebrated as progressive, it also entails psychological trauma.

One of the challenges in creating Soft Cinema films is to come up with narratives that have a structural relationship to the database aesthetics. If Texas uses semi-random database retrieval to represent 'info-subjectivity', then *Mission to Earth* adopts the variable choices and multi-frame layout of the Soft Cinema system to represent 'variable identity'. That is, the trauma of immigration, the sense of living parallel lives, the feeling of being split between different realities. To this end, in generating every part of the film, the software chooses from among a number of alternative sequences that reflect Inga's variable identity. Other factors, such as the choice of a large or smaller window to display a particular sequence, and the number of windows (co-present realities) that appear in a layout, simultaneously tell us what the main character is seeing and represent her thoughts, memories and feelings.

One of the goals of this film was to visualize the narrative as much through motion graphics as through live action video. Consequently we invited Ross Cooper Studios to create a database of short motion graphics clips, which would respond to the film's narrative and to the live video footage. In most parts of the film you will see both video clips and motion graphics clips appearing side by side. The motion graphics react to the video but they also hold their own. In fact they form a parallel film that follows the same narrative but visualizes its themes and the feelings of the characters through different means.

MISSION TO EARTH

CREDITS

Mission to Earth (2003–2004) was commissioned and produced by BALTIC The
Centre for Contemporary Art (Gateshead, UK).
[Lev Manovich | narrative, videography, editing] [Kelly Richardson | Newcastle |
assistant director & editor] [CKUK | Christopher Kent | London | narrator]
[Ilze Black | London | playing Inga] [Alec Finlay | Newcastle | playing Alpha-1 com-
mander] [Jóhann Jóhannsson | Iceland | music] [servo | Los Angeles, New York,
Zurich, Stockholm | architectural designs] [Ross Cooper Studios | Ross Cooper &
Stuart Sinclair | London | motion graphics   Martins Ratniks | Ernest Karlsons |
Ilze Black | Latvia | videography] [Christo Wallers | Stuart Harris | Newcastle | vid-
eography] [Tom Cullen | Newcastle | technical coordination]

Ilze Black / Inga

Born in Riga (Latvia) Ilze Black is an artist, curator, writer, a co-founder of Ambient Information Systems, and a research fellow at Goldsmiths College, University of London. As a founding member of the Latvian-based art bureau OPEN, since the mid-1990s she has staged such events as *Untitled: subvertizing session in the streets of Riga*, and *T-shroom*. Most recently her work has been developed inside the framework of the Trans-European Re_Public Art network. Projects include a 24-hour live online road movie titled *Broadbandit Highway*, and the network architecture project *Ambient Wireless*.

1

2

3

4

1   Newcastle
2|6   Tokyo
3   Irbene Radio Telescope, Latvia
4   Rotterdam
5   studio set

Irbene Radio Telescope /ALPHA-1 Space Center

Irbene Radio Telescope (Latvia) is a 32 meter-wide, computer controlled, fully steerable, parabolic dish that became operational in 1971. While its exact purpose remains secret, it was most probably used to spy on satellite transmissions between Europe and North America. Abandoned and nearly destroyed when the Russian Army departed Latvia in 1994, the dish has been successfully repaired by the Ventspils International Radio Astronomy Center (VIRAC). Since 2001 the Riga-based new media center RIXC has organized a number of international art workshops and events that utilize the radio telescope.

1

1 main telescope dish
2|3 top observation area
4 control room

2   3   4

1

ALPHA-1

main sources:

2

3   4   5

6   7   8

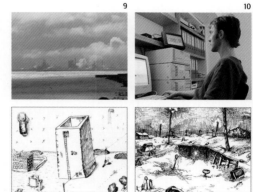

9   10

11   12

9   the industrial area in UK which inspired
Ridley Scott's *Blade Runner* (2003 video)
10   Baltic Centre control room (UK, 2003)
11 | 12   drawings by Lev Manovich (1981–1991)

My idea for this Soft Cinema edition was to create a film without a pre-formulated narrative. I started with two assumptions. Firstly that meaning and narrative coherence can be created through associative chains. Secondly – presuming that human perception works to connect images that are presented together and to integrate them into a coherent, meaningful, structure – that it is possible to create these associative chains through the visual properties of images.

Absences explores two narrative fields, which can be described as the field of urban landscape and the field of natural landscape. Each field is represented by a number of film fragments and each fragment is described via a number of parameters that capture its pictorial properties. I went through a process of rearranging, associating, and adjusting the parameter space until the parameters adequately captured various visual distinctions between the clips, such as the distinction between concrete and abstract.

As I worked with the database and the parameter space, the idea for a narrative came to me. And, as I made editing decisions and developed the two parallel threads that are intermittently presented throughout the project, so my narrative idea solidified into the structure that now guides the viewer through the film. On the one hand, this structure is an investigation of the pictorial and kinetic dimensions of images. On the other hand, it describes the quest of the narrator-ego (represented by the camera) for a peaceful and pristine paradise. Throughout an ever more frantic quest in the labyrinth of a city, the ideas of paradise and reality become increasingly confused and eventually fuse into abstraction.

Soft Cinema software does not generate an aesthetic result per se. The software was used as an associative tool that enabled me to explore a collection of film clips and to arrange them in a coherent way. The machine processed the material without any aesthetic preconception and this allowed for new narrative and aesthetic structures to arise from an initially indiscriminate database. So, while it is in the end an algorithm that tells the Soft Cinema display software to show a certain sequence of video clips, the algorithm itself is the result of an authoring process. It is, consequently, only through my creative decision-making – regarding which clips to include in the database, which parameters to select, how to weight them, and which rhythm in the temporal development to follow – that the final film takes on aesthetic qualities.

5   6

Jóhann Jóhannsson /music

Jóhann Jóhannsson is best known as one of the founders of Kitchen Motors, the Reykjavik based art/music think-tank and record label and as a member of the Apparat Organ Quartet, which has performed widely in Europe and the US. He has worked with such internationally known artists as Barry Adamson, Pan Sonic, Marc Almond, and the Hafler Trio and composed music for several films as well as for theatrical and dance productions. In 2002 Jóhannsson's first solo album *Englabörn* – tracks from which are used in *Mission to Earth* – was released on the UK based Touch label.

▸ www.kitchenmotors.com

Kelly Richardson /assistant director & video editor

Currently living and working in England, artist Kelly Richardson was born in Canada where she studied fine art at the Ontario College of Art & Design and the Nova Scotia College of Art and Design. Her work focuses on the restitution of the sublime from an ordinary or flawed moment and her projects have been exhibited at various notable international venues, most recently the Gwangju Biennale, South Korea; Art Gallery of Nova Scotia, Canada; Cornerhouse, England; and Stills Gallery / Edinburgh Film Festival, Scotland.

▸ www.kellyrichardson.net

servo /architectural designs

Established in 1999, servo's experiments with emergent design, fabrication, and interactive information technologies focus on the complex interface of emerging media and architectural practice. Comprising an international network that links the architecture cultures of Europe and the United States, servo has held six solo exhibitions, participated in eleven group exhibitions, and its members teach and lecture widely on both continents.

Decentralized across four cities, the collaborative borrows its name from the servo motor, an apparatus common in the field of cybernetics. Servo motors translate digital code into machinic processes. They behave principally as enablers, allowing two discrete languages to converse and interact. In a similar way servo organizes, coordinates, and ultimately enables new and existing relations between technologies, forms of disciplinary expertise and modes of production, as well as between a variety of cultural influences that are specific to the cities from which it operates.

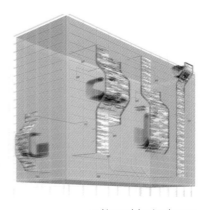

architectural drawings by servo

Ross Cooper Studios /motion graphics

Ross Cooper is a designer and consultant for many companies, including the BBC's Creative Research & Development Department and Ford's Future Products Group. As a media artist his works have been exhibited at Ars Electronica, Linz; Kiasma, Finland; The Digital Hub, Ireland; Gifu Prefecture Takumi Studio, Japan; Teatro Miela, Italy and El Ciutat, Barcelona. His awards include an Honorary Mention from Prix Ars Electronica, 2 Silver D&AD's and a Gold from the Art Director's Club of Europe.

Designer Stuart Sinclair works in the areas of moving image and interaction design. His projects include film and television title sequences produced with Fernando Gutierrez, a partner at Pentagram, London and Princess Productions, London.

▸ www.rcstudios.com

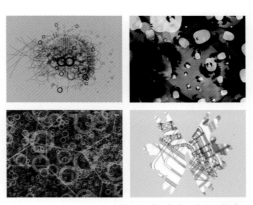

motion graphics by Ross Cooper Studios

Montage

In addition to investigating the ways in which algorithmic associations between visual elements and narrative can work together, *Absences* re-examines several aspects of montage. Although the idea of spatial montage emerged rather early in film history, it was realized only in a few avant-garde films. Recently, however, it has become commonly used in the medium of television with many programs presenting several independent streams of information simultaneously on the screen.

Frames from clips used in
ABSENCES
(shot in Split and Berlin)

ABSENCES

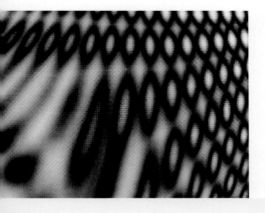

CREDITS

[Andreas Kratky | Berlin
videography, music, editing, modified version of Soft Cinema software]

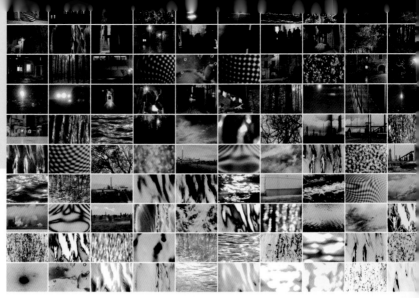

database sorted by contrast

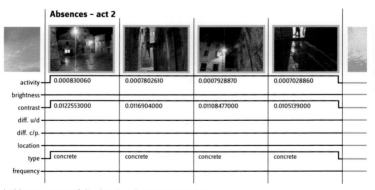

Absences – act 2

activity	0.000830060	0.0007802610	0.0007928870	0.0007028860	
brightness					
contrast	0.0122553000	0.0116904000	0.01108477000	0.0105139000	
diff. u/d					
diff. c/p.					
location					
type	concrete	concrete	concrete	concrete	
frequency					

building a sequence of clips based on their parameter values

Parameters

For the *Absences* database each film clip was analyzed in regards to several visual qualities: brightness, contrast, texture, activity, frequency, the difference between the center and the periphery, and the difference between the upper and lower parts of a frame. All of these parameters were described using numerical values that were acquired with image processing software. In addition I have also introduced a manually defined parameter that allows for differentiation between concrete and abstract clips. By combining these various parameters and assigning different weights to them, the clips were organized and diverse modes of associating them were developed.

database sorted by activity

Christine Bokelmann /graphic design

Berlin-based graphic designer Christine Bokelmann has been responsible for translating the ideas of Soft Cinema into a print form. To design this booklet, as well as the 2002 Soft Cinema catalog, she used Soft Cinema software to generate both the 2D grids and the still images. These elements became 'building blocks' out of which the documents were constructed.

Janet Owen /Soft Cinema DVD production manager, catalog editor

Writer, curator and co-founder of the AIM festival, Janet Owen, has organized exhibitions and related events with, among others, MOCA, LA; Kiasma, Finland; the Hong Kong Art Center; Tsinghua University, Beijing; and the Susquehanna Art Museum, US. Her texts have appeared in various exhibition catalogs; magazines, including *RIM* and *Le Rouge*; and the essay anthology *Kolibri* (Revolver, 2003).

Absences uses up to five streams of video. In order to facilitate the reading of these streams, one stream is always made large and thereby prioritized over the accompanying subordinate streams. (This is the general aesthetic strategy that is built into Soft Cinema software). The smaller image streams adopt different roles throughout the film: they are synchronous additional perspectives to the main stream; they are small glimpses into the future narrative of the film; or they reflect what has already passed. The existence of such parallel streams raises the interesting question of density – how much information can the viewer deal with given the nature of the footage and the speed of the narrative?

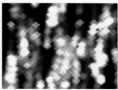

Frames from clips used in
ABSENCES
(shot in Newcastle area and Berlin)

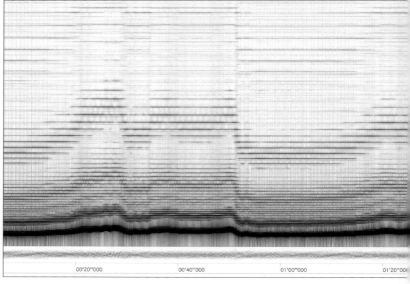

Spectra of sound layers from ABSENCES

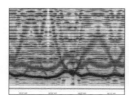
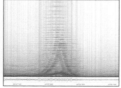

Sound

The soundtrack of *Absences* comes from the same parameter space as is used to describe the visual clips. The curves describing the temporal development of the activity and the brightness and contrast values of the individual clips were used to synthesize sounds, which were then combined with both location sound and original compositions. The result is a sound track that loosely follows the images. As several clips appear on screen, so their accompanying sounds become layered. In this way the rhythm of the filmic montage also creates a sonic rhythm. The overall film sound track follows the progression of the visuals from more concrete to more abstract by moving from location sound to layered sound.

Translation from image parameters into sound

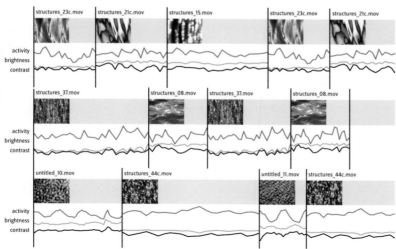

Our research follows four directions:

1. Following the standard convention of the human-computer interface, the display area is always divided into multiple frames.
2. Using a set of rules defined by the authors, the Soft Cinema software controls both the layout of the screen (number and position of frames) and the sequences of media elements that appear in these frames.
3. The media elements (video clips, sound, still images, text, etc.) are selected from a large database to construct a potentially unlimited number of different films.
4. In Soft Cinema 'films' video is used as only one type of representation among others: motion graphics, 3D animations, diagrams, etc.

Together these directions define a new aesthetic territory. The three films presented on the Soft Cinema DVD explore some parts of this terrain.

When they are shown as installations, each of the films is assembled by the Soft Cinema software in real time. As a result a film can run indefinitely without ever exactly repeating the same edits. To adapt the films to the DVD medium we capture specific software 'performances' directly off the screen. All these alternative versions are placed on the DVD, which is programmed to navigate between them. Consequently there is no single 'unique' version of each film. Not everything will be different with every viewing, but potentially every dimension of a film can change, including the screen layout, the configuration and combination of the visuals, the music, and the narrative.

The following pages introduce these films and the people who worked on them in more detail.

LEV MANOVICH